# The United States Army War College

The United States Army War College educates and develops leaders for service at the strategic level while advancing knowledge in the global application of Landpower.

The purpose of the United States Army War College is to produce graduates who are skilled critical thinkers and complex problem solvers. Concurrently, it is our duty to the U.S. Army to also act as a "think factory" for commanders and civilian leaders at the strategic level worldwide and routinely engage in discourse and debate concerning the role of ground forces in achieving national security objectives.

SSI
STRATEGIC STUDIES INSTITUTE

The Strategic Studies Institute publishes national security and strategic research and analysis to influence policy debate and bridge the gap between military and academia.

CENTER for STRATEGIC LEADERSHIP CSL

The Center for Strategic Leadership contributes to the education of world class senior leaders, develops expert knowledge, and provides solutions to strategic Army issues affecting the national security community.

PKSOI

The Peacekeeping and Stability Operations Institute provides subject matter expertise, technical review, and writing expertise to agencies that develop stability operations concepts and doctrines.

SCHOOL OF STRATEGIC LANDPOWER
U. S. Army War College

The School of Strategic Landpower develops strategic leaders by providing a strong foundation of wisdom grounded in mastery of the profession of arms, and by serving as a crucible for educating future leaders in the analysis, evaluation, and refinement of professional expertise in war, strategy, operations, national security, resource management, and responsible command.

The U.S. Army Heritage and Education Center acquires, conserves, and exhibits historical materials for use to support the U.S. Army, educate an international audience, and honor Soldiers—past and present.

# STRATEGIC STUDIES INSTITUTE

The Strategic Studies Institute (SSI) is part of the U.S. Army War College and is the strategic-level study agent for issues related to national security and military strategy with emphasis on geostrategic analysis.

The mission of SSI is to use independent analysis to conduct strategic studies that develop policy recommendations on:

- Strategy, planning, and policy for joint and combined employment of military forces;
- Regional strategic appraisals;
- The nature of land warfare;
- Matters affecting the Army's future;
- The concepts, philosophy, and theory of strategy; and,
- Other issues of importance to the leadership of the Army.

Studies produced by civilian and military analysts concern topics having strategic implications for the Army, the Department of Defense, and the larger national security community.

In addition to its studies, SSI publishes special reports on topics of special or immediate interest. These include edited proceedings of conferences and topically oriented roundtables, expanded trip reports, and quick-reaction responses to senior Army leaders.

The Institute provides a valuable analytical capability within the Army to address strategic and other issues in support of Army participation in national security policy formulation.

**Strategic Studies Institute**
**and**
**U.S. Army War College Press**

# SOCIAL MEDIA—THE VITAL GROUND: CAN WE HOLD IT?

**Ian Tunnicliffe**
**Steve Tatham**

**April 2017**

The views expressed in this report are those of the authors and do not necessarily reflect the official policy or position of the Department of the Army, the Department of Defense, or the U.S. Government. Authors of Strategic Studies Institute (SSI) and U.S. Army War College (USAWC) Press publications enjoy full academic freedom, provided they do not disclose classified information, jeopardize operations security, or misrepresent official U.S. policy. Such academic freedom empowers them to offer new and sometimes controversial perspectives in the interest of furthering debate on key issues. This report is cleared for public release; distribution is unlimited.

*****

*****

Comments pertaining to this report are invited and should be forwarded to: Director, Strategic Studies Institute and U.S. Army War College Press, U.S. Army War College, 47 Ashburn Drive, Carlisle, PA 17013-5010.

*****

This manuscript was funded by the U.S. Army War College External Research Associates Program. Information on this program is available on our website, *ssi.armywarcollege.edu,* at the Opportunities tab.

*****

All Strategic Studies Institute (SSI) and U.S. Army War College (USAWC) Press publications may be downloaded free of charge from the SSI website. Hard copies of certain reports may also be obtained free of charge while supplies last by placing an order on the SSI website. Check the website for availability. SSI publications may be quoted or reprinted in part or in full with permission and appropriate credit given to the U.S. Army Strategic Studies Institute and U.S. Army War College Press, U.S. Army War College, Carlisle, PA. Contact SSI by visiting our website at the following address: *ssi.armywarcollege.edu.*

*****

The Strategic Studies Institute and U.S. Army War College Press publishes a quarterly email newsletter to update the national security community on the research of our analysts, recent and forthcoming publications, and upcoming conferences sponsored by the Institute. Each newsletter also provides a strategic commentary by one of our research analysts. If you are interested in receiving this newsletter, please subscribe on the SSI website at the following address: *ssi.armywarcollege.edu/newsletter/.*

ISBN 1-58487-752-9

## FOREWORD

In this timely and realistic examination of social media, two world-class British experts examine exactly, in the defense context, what social media is and what it should and should not be used for in the future. In setting out their arguments, they define social media in four distinctly different ways: first, they assess, perhaps self-evidently, that it is a media channel, and actually differs little to newspapers and radio in anything other than reach and immediacy; second, they see it as an interactive medium that might have potential for exerting influence, but only when accompanied by robust target audience analysis (TAA); third, they see it as a means of establishing a dialogue and communications within already well-established networks and groups; and, finally, they see it as a real-time sensor network that may possibly provide the first indication of globally important events—albeit unsubstantiated and raw in its content and reporting. They also make a clear statement; the evidence that social media, by itself, as a precursor to mass behavior is not present. Social media, in the authors' views, cannot predict behaviors with any reliability or consistency.

Having considered each of these in the military context, the authors then examine the military implications of social media. Here they address four questions: First, is social media a viable and effective messaging conduit? Second, what are the implications of social media for information operations (IO) doctrine, personnel, and operational security (OPSEC)? Third, the authors ask if the United States has the necessary structures, training, and equipment in place to effectively harness social media. Finally, the authors ask how well our institutions are placed to train and educate military personnel—and specifically the most senior personnel—in what social media can and cannot achieve.

Social media, the authors write, is a disruptive, yet innovative, technology that is poorly understood by the military, and it does not sit well in current hierarchical military structures. In headquarters and operations rooms, globally, the Internet terminal is typically the preserve of the public affairs officer (PAO). The mixing of open source Internet-enabled IT systems and classified military and governmental communication systems is fraught with difficulty—bureaucratic and technological. If the military is to leverage the full capability of social media, there will be many "sacred cows" that will have to be slain. However, commanders, the authors conclude, should ignore them at their own peril.

Throughout this Letort Paper, the authors draw upon their vast IO experience gathered from multiple global operations to present a fascinating and useable insight into a phenomena that is not yet 15 years old.

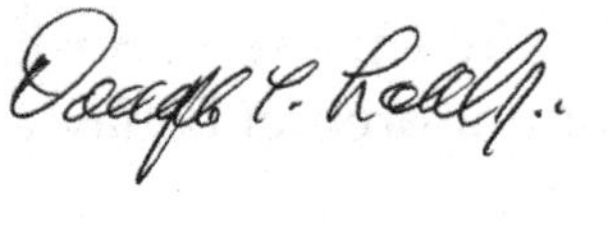

DOUGLAS C. LOVELACE, JR.
Director
Strategic Studies Institute and
U.S. Army War College Press

## ABOUT THE AUTHORS

IAN TUNNICLIFFE is a Director at Accordance Associates and has led communications, research, and training projects across the Middle East, Africa, and Asia, including Iraq, Libya, Egypt, Nigeria, Somalia, and Afghanistan. A visiting lecturer in social media to the North Atlantic Treaty Organization (NATO) School in Oberammergau, Germany, his recent work has focused on the role of social media in conflict areas, in particular in Syria, Nigeria, and Somalia. He has also continued to provide the United Kingdom (UK) with exercise support for the Allied Rapid Response Corps (ARRC), serving as the Head of Social Media in Exercise Trident Jaguar 2015 and Exercise Arrcade Fusion 2015. Prior to that, he served for 20 years in the British Army, rising to the rank of Colonel, and his last few years of service were with the Directorate of Targeting and Information Operations (DTIO) in the UK Ministry of Defence (MOD), where he helped develop UK cross-government strategic communications plans in response to a number of international crises in the Balkans, Africa, Asia, and especially the Middle East during the Iraq crisis and conflict. His last post in the Army was in Baghdad, Iraq, during the period between September 2003 to June 2004, where he worked as Director Plans in the Office of Strategic Communications of the Coalition Provisional Authority (CPA), working for Ambassador Paul Bremer. During his tenure, he was responsible for Iraq-wide Coalition and United Nations (UN) media campaigns to inform the population of Coalition and UN activities.

STEVE TATHAM is a private consultant specializing in Strategic Communication, influence, target audience analysis, and information operations. Dr. Tatham resigned from the UK's Armed Forces in July 2014. On leaving, he was the UK's longest continuously serving officer in information activities. Between 1998 and 2003, he worked in media operations, covering conflicts in Sierra Leone (2000), Afghanistan (2001), and Iraq (2003), where he was a public spokesperson for the invasion. Between 2003 and 2013, he worked in information operations and psychological operations (PSYOPS). Dr. Tatham was the Commanding Officer of 15 (UK) PSYOPS Group from 2010 to 2013, during which time he deployed on multiple occasions to Afghanistan; was involved at the operational level in operations in Libya; and deployed to East Africa in an advisory role. His final military appointment was as the Special Information Operations Project Officer in the UK Ministry of Defence Operations Directorate. In 2007, he advised the then-commander of British Forces in Afghanistan on influence operations when the strategically vital town of Musa Qala was retaken by British and Afghan forces. The Pentagon later described that operation as the: "single best thing to come out of Afghanistan." Dr. Tatham is the author of two books: *Losing Arab Hearts & Minds: The Coalition, Al-Jazeera and Muslim Public Opinion* (Hurst & Co, 2006); and, *Behavioural Conflict: Why Understanding People's Motivations Will Prove Decisive In Future Conflict* (Military Studies Press, 2012). Dr. Tatham holds an M.Phil. and a Ph.D. in international relations, both focusing on ideas of influence and strategic communication in conflict areas.

## SUMMARY

This Letort Paper seeks to answer some specific questions on how the U.S. Army, and by extension its allies, can best leverage social media, particularly on operations. Are they a viable and effective messaging conduit? Have they changed the information operations (IO) landscape? Does the United States have the correct force structure, training, equipment, and technology to leverage social media? Finally, how can we train our leaders to better understand and use social media?

The impact of social media on the media environment has been widely recognized, as has the ability of extremist and adversarial organizations to exploit the media to publicize their cause, spread their propaganda, and recruit vulnerable individuals. Supporting the growth of social media has been the phenomenal global increase in mobile telephone usage, and much of this increase is in areas where there are existing conflicts or conflicts are highly likely.

However, the full implications of the mobile and social media revolution are not yet fully understood. Social media will increasingly have a direct impact on virtually all aspects of military operations in the 21st century. In doing so, social media will force significant changes to policy, doctrine, force structures, and virtually all staff functions within operational units. New training requirements and new approaches to traditional operational challenges will be required. All this will require a wider understanding of social media and the realization that it is no longer an area simply of concern to public affairs officers (PAO) and possibly intelligence (INT).

Although with new threats come new opportunities, and the interactive nature of social media in particular means that it is potentially a very powerful medium for IO, that potential needs to be recognized before capabilities can be developed to respond to this rapidly developing revolution.

There are four distinctly different applications of social media; and understanding each, together with its impact upon the Army, is critical to addressing the various questions posed.

- Social media can be regarded primarily as a media channel, just like radio, newspapers, and television.
- In addition to this, social media can be seen as an interactive medium for exerting influence.
- Importantly, social media is a way of communicating within an already established network or networks.
- Finally, social media is a near real-time sensor-to-sensor network.

There is a need for commanders, and in particular operations and INT staff, to understand social media and all its different functions, not simply its media function, in order to understand its potential impact on operations, and to incorporate that understanding into their planning and operations.

The key recommendations include:

- A social media capacity needs to be built into every level of command. In the real-time information environment created by social media, operational commanders at every level will need simultaneous and identical situational awareness of unfolding events.

- Social media is integral to the conduct of operations, not just a PAO or INT function. Thus, the U.S. Army needs a custom-made doctrine and an educational capacity to inform commanders on its intelligent and safe use.
- In order to integrate social media into operations, each of the staff functions needs continuous access, at their desk, to the Internet.
- Social media must be integral to all exercises. This needs the creation of a virtual social media environment—i.e., a sandbox or simulation—to add the social media space to training environments. This must be a space in which it is "safe to fail."
- Within headquarters, the responsibility for social media needs to be clearly articulated and the posts resourced appropriately.
- U.S. Army and Department of Defense (DoD) policy and doctrine must clarify a host of difficult issues relating to social media usage, in particular its use in deception and Psychological Operations (PSYOPS).

## SOCIAL MEDIA—THE VITAL GROUND: CAN WE HOLD IT?

### INTRODUCTION

Once upon a time, there existed a world without social media. In the ubiquitous information environment in which we now live, it seems incredible that such a world was, in fact, no more than 12 years ago. Between 2004 and 2014, all 22 of the world's biggest social media networks were developed and launched, with the first, Facebook, on February 4, 2004.[1] Supporting that growth has been the phenomenal global increase in mobile telephone usage. While this was initially driven by the consumerism of the global "West," today it would be a challenge to find a society where there is no mobile phone usage.[2] Much of the rapid increase is concentrated in areas where there are existing conflicts, or conflicts are highly likely, and as a result, we believe that mobile phone usage, particularly when associated with social media, is likely to have a significant impact upon the conduct and outcome of those conflicts.[3] However, can we quantify that outcome? Much seems to have been written about social media: the apparently extraordinary recruiting power of the so-called Islamic State of Iraq and Syria's (commonly referred to as ISIS) social media campaign (of which we remain to be convinced); the apparent "success" of Russia's trolling (of which we are absolutely less convinced); and the way that social media apparently triggered the Arab Spring (of which we are absolutely not convinced at all).

It may actually be too late to undertake a nuanced assessment of the challenge of social media from the perspective of policy—many governments and militaries have apparently already decided that the social media space is a battlefield and are already spending millions on creating social media soldiers or "network warriors" to engage in those online battles. Whether that is money well spent is debatable, but regardless, in this Letort Paper we seek to answer some specific questions on how the U.S. Army, and by extension its allies, can best leverage social media, particularly on operations. Specifically, we will consider if social media is really a viable and effective messaging conduit; secondly, we investigate if social media has changed the information operations (IO) landscape; thirdly, we will consider if the United States has the correct force structure, training, equipment, and technology to leverage social media. Finally, we will consider how we train and educate our leaders to better understand and use social media.

Before addressing each of these areas, we need to contextualize our thoughts with our observations of what social media is, or can be, in the military context. We find four distinctly different applications of social media, and understanding each, together with its impact upon the Army, is critical to addressing the various questions posed.

1. Social media, in our view, should be regarded primarily as a media channel, just like radio, newspapers, and television.
2. In addition to this, social media should be seen as an interactive medium for exerting influence.
3. Importantly, social media is a way of communicating within an already established network or networks.[4]
4. Finally, social media is a near real-time sensor-to-sensor network.[5]

Each of these observations needs to be explained in a little more detail.

The first function, social media as a media channel, is perhaps what people most clearly understand. The spread of social media has transformed the speed with which incidents are reported, and in many cases, events are now being reported through social media as they are still unfolding.[6] This has undoubtedly changed the nature of news reporting and has shortened the response time available to government and military leaders after major events. We need to caveat this, however, with some historical context. This response time has been reducing throughout history with the advent of new communications technologies—with steep changes at the introduction of significant innovations such as the post horse and telegraph networks—so this should not be regarded as a new phenomenon. However, social media as a media channel has enabled policymakers to see a wider range of viewpoints than ever before, including those of the adversary. It has also created a dilemma: most conventional news reporting follows certain standards of truth and accuracy. The British Broadcasting Corporation (BBC), for example, will only report an event as news if it has two independently verifiable sources (unless a BBC correspondent is witnessing the actual event).[7] If it does not, the BBC will clearly state that the report has yet to be verified. Such moral exactitude does not exist in social media, so policymakers are gaining a far wider view of an event, but the accuracy of that view is highly debatable. While this is a major challenge, the nature of the challenge is, in general, well understood, especially by public affairs professionals.[8]

Unlike conventional media, though, social media allows individuals to interact with others in multiple different ways and directions. It is this second function that enables social media to be considered as a potentially powerful medium for influence. Much recent military and governmental investment in social media has been in using it to spread counter-narratives in response to the all too frequently successful use by extremist groups. The West has looked at the use of social media by organizations such as ISIS and concluded that this is a critical area whose importance will only increase. We agree that this is an area of growing importance; communities currently without social media penetration can reasonably be expected to, in time, exhibit the same characteristics of communities that already make use of it today. However, there is also a very real danger in assuming if social media says it is so, it is so. There is also a presumption that the volume of social media postings is indicative of interest. Again, we urge caution—volume can be manufactured very easily.[9] Overly focusing on this area, while important, risks overlooking the other functions of social media that we have identified in this Letort Paper previously.

We believe that insufficient attention has been paid to the nature of social media as a communications network *per se*. It is known that terrorist and insurgent groups already use applications such as SnapChat and WhatsApp for tactical communication, but this is just the tip of the iceberg, because they are now increasingly being used by military and civilian groups of all descriptions.[10] For example, a colleague and European North Atlantic Treaty Organization (NATO) officer who was deployed to sub-Saharan Africa regularly used WhatsApp to communicate with colleagues in different European nations, seeking advice on operational issues. The officer explained that the app was almost instantaneous, that it could reach targeted people whose advice was trusted, and was far faster than any military information technology (IT) system, which would not have been able to reach specialists of different nationalities at all, let alone in the time-frame that the data was needed. A second example is the wave of immigrants making their journey to Europe. A number of studies have now shown that WhatsApp, in particular, is being used as a tool to improve the immigrants' situational awareness.[11]

Although messaging apps like these have been seen as less secure than government communications and, therefore, more open to interception, even this may be changing. The recent court case between the U.S. Government and Apple over the decryption of an iPhone suggests that commercial security protocols are becoming at least as strong as many governmental ones.[12] Whether less secure or not, they are all more user-friendly and accessible than conventional classified systems. While the U.S.'s Secret Internet Protocol Router Network (SIPR) and Non-classified Internet Protocol Router (NIPR) network systems are relatively sophisticated (noting that it is very difficult for non-U.S. nationals to gain access), the United Kingdom's (UK) Defence Information Infrastructure (DII) system, for example, is notorious for its slowness, its lack of connectivity with others, and its questionable "user friendliness." We should, therefore, not be surprised when digitally aware officers use their initiative and the technology they are familiar with to address issues quickly and safely. It seems to us that in the future, social media platforms, often the same ones, will inevitably be used for operational communication by our people, by local civilian populations, and by adversarial groups alike.

The final function we would highlight is the ability to use a network of smartphones effectively as a sensor network. A population equipped with smartphones and willing to communicate to others about the events taking place in their area is able to generate a picture of ongoing events in real time across the span of that population. Through the use of techniques such as crowd sourcing—the practice of obtaining information by soliciting contributions from a large group of people on social media—civil society organizations (CSOs) are already exploiting this capability in humanitarian crisis situations.[13] To date, crowd sourcing has been used to generate situational awareness for humanitarian organizations in disaster zones, from the aftermath of typhoons in the Philippines, to the earthquake in Haiti, to major flooding in Australia. In each case, an accurate picture of the situation was built up virtually and almost entirely through social media posts and updates across the area concerned.[14]

What social media cannot do yet, in our view, is predict behavior with any reliability or consistency. This latter finding will be the most contentious, not least, as many commercial companies are actively marketing their media monitoring tools as behavioral predictors. Undeniably, while terrorists use the web to communicate—particularly applications that do not place their data sets in the public domain such as WhatsApp—the evidence that social media, by itself, is a precursor to mass behavior is simply not present. As the U.S. Peace Institute concluded in a major report on the Arab Spring: "**New media... did not appear to play a significant role in either in-country collective action or regional diffusion during this period** [emphasis added]." In any event, the Arab Spring, such is the pace of social media development, is already of limited relevance for assessing current capabilities.[15]

It would also be useful at this early stage to identify some of the constraints and challenges social media poses. For example, social media is a disruptive, yet innovative technology that does not sit well in current hierarchical military structures. In headquarters and operation rooms globally, the Internet terminal is typically the preserve of the public affairs officer (PAO). The mixing of open source Internet-enabled IT systems and classified military and governmental communication systems is fraught with difficulty—bureaucratic and technological. If the military is to leverage the full capability of social media, many sacred cows will have to be slain. Social media is not just a public affairs or strategic communication-messaging tool; it is already a major source of intelligence, and will become more important still in understanding and planning military operations. Commanders ignore it at their peril.

## HOW DOES THE ARMY LEVERAGE SOCIAL MEDIA?

How can the Army leverage social media for strategic messaging? First and foremost, the Army needs to understand the changing dynamics and demographics of social media across the world and, in particular, the way in which people now access information. The world is in the middle of a so-called "mobile revolution," and increasingly, the principal way to reach external audiences worldwide is through social media via their mobile. In the context of the likely audiences in conflict environments, this trend is even more pronounced, as these environments are often in less developed countries where the population has previously had very limited access to the Internet.

The evidence indicates that the acceptance of data-enabled smartphones has been even faster than previous revolutions in communications such as radio.[16] Mobile data traffic has grown 4,000-fold over the past 10 years, and the speed of this device acceptance continues to increase; global mobile data traffic grew 74 percent in 2015.[17] The world is also nearing the so-called "peak device" point, as it is estimated that more than half of all Internet traffic will originate with non-personal computer devices by 2019.[18] This trend is predicted to continue, and mobile data traffic is predicted to grow at a compound annual growth rate (CAGR) of 53 percent from 2015 to 2020. By then, it is estimated that there will be 11.6 billion mobile-connected devices, exceeding the world's projected population at that time (7.8 billion).[19]

Social media is already the primary use for the data capability of smartphones. In 2015, there were 3.42 billion Internet users (equaling 46 percent global penetration), but not that far behind, there were 2.31 billion social media users (31 percent global penetration).[20] The impact of the mobile revolution is worldwide, but the areas that are most dramatically affected by these changes, with the highest rates of smartphone and Internet growth, are Africa, the Middle East, and Central Asia, including many fragile states and areas where conflict is ongoing or where the potential for conflict exists.[21] While television and radio remain important sources of information, the growth of social media has seen it become an increasingly trusted source, especially likely to be trusted by the youth demographic.[22]

In addition to accessing a range of news sources, including both national and international media channels, social media enables access to local sources of information such as bloggers and local websites, and CSOs working in the local area. The news can cover anything: the locations of current fighting, the state of repair of power and water lines, the condition of roads and the possibility of travel, the names of those killed in a recent strike, food prices and locations, international news, and media statements made by politicians, activists, or military leaders.

Perhaps most importantly, social media enables contacts with friends and family, traditionally the most trusted sources for information in many societies, particularly in developing countries. In current conflict zones such as Syria, communications within such networks have been greatly enhanced by the use of social media. In his description of the importance of Facebook in Syria, blogger Ammar Halabi writes:

> Browsing the Facebook feed has therefore [become] a primary way to get news and commentary about the whereabouts of friends and family, either inside or outside of Syria. My connections have mentioned often that browsing the 'green dot' that indicates that somebody is online on Facebook chat was a way to know that they are doing OK [emphasis added].[23]

While Facebook, WhatsApp, and Twitter have come to dominate across all demographics in many countries where social media is in its infancy, the increasing maturity of the communications environment in other countries has led to greater audience segmentation, because different social media platforms have become more popular with different sections of the population. In the UK for example, the average age of a Facebook user is increasing, and the majority of users are now in the 25 to 34 demographic, while applications such as Snapchat are primarily used by 18-24 year olds and Kik by an even younger demographic.[24] This trend is only likely to continue and will increasingly allow better-targeted communications with specific audience segments.

With the continuing evolution of social media, many countries are developing specific preferences for social media applications divided on demographic, political, or even ethnic lines. For example, in Latvia most of the population use the Latvian "Draugiem" social media site, with limited penetration of the primarily Russian social networking site VKontakte (VK), suggesting a political division of social media between those considering themselves to be Latvian Latvians and those Latvians of Russian descent.[25] Similarly, suspicions that Russia was using data from VK to spy on Ukrainians resulted in an increased use of Facebook by those opposed to Russian operations in Ukraine.[26]

The differentiation of social media in different countries also extends to specific uses. WhatsApp is popular across the world, but primarily as a messaging app; it is frequently used as a free substitute for texting. In Sudan, however, activists are using it to share news and discussion.[27] With the repressive government in Sudan monitoring other social media, such as Facebook, activists prize WhatsApp for its security and encryption, and through the use of message-groups activists have adapted it to widely distribute news.

The mobile revolution and phenomenal rise of social media have affected virtually every country in the world, including countries assumed to lack any significant Internet penetration at all. In Afghanistan, for example, the percentage of population with access to the Internet has risen to more than 12 percent in a few short years, and that percentage continues to rise at a rapid pace. It now has over 4 million Internet users, of whom 2.6 million are on Facebook.[28] Its increasing importance was highlighted in December 2015, when the deputy governor of the Helmand province sent an open letter to the President pleading for more assistance in the province—by means of a Facebook update.[29] Paradoxically, President Erdoğan, no lover of social media in Turkey, found it unexpectedly useful to communicate to his people in the attempted July 2016 coup.[30] In short, in the near future there is no potential conflict area in the world that will not be significantly impacted by the Internet and social media.

Collectively these trends clearly indicate that careful target audience analysis (TAA) of the social media environment needs to be carried out to determine what channels to use. They also suggest that targeted messaging to specific demographics is increasingly feasible. However, the issue of credibility acts as a major constraint. Recent years have seen a reduction in the assessed trustworthiness of government communications, particularly for certain demographics such as youth and minority groups.[31] It is these groups that are frequently key audiences for strategic messaging. Where the Army is attempting to communicate, the interactive nature of social media can provide better ways of engaging audiences, particularly younger audiences, but the emphasis needs to be on creating an informal, open, and interactive approach, along with utilizing credible non-government voices wherever possible.

In addressing the question of how the Army leverages social media, the primary lesson is that the Army needs to have a better understanding of its audiences, and their use of social media, than it does at present. Both authors have worked within the UK, U.S., and coalition commands and have shared discussions with uniformed colleagues of all nationalities. The paucity of social media understanding and access has been a recurring theme of debate. What this means, in practice, is allowing commanders and planners a deeper and broader understanding of the specific demographics and usage of social media in their respective regions of interest. This may seem a *non sequitur,* but our experience is that many senior, and certainly older, people tend not to understand the details of social media, employing instead a "one size fits all" approach. Secondly, the Army needs to understand usage patterns. Again, this is largely a function of TAA; it is no good placing all of the output on Facebook if the majority of users are on VK. Inherently, this means that the Army needs strong linguistic skills in order to understand and seamlessly slip into these social media outlets.

Thirdly, the Army needs to have a good understanding of the political and social divisions that characterize social media usage. In Latvia, for example, we found that many senior people made the assumption that Russian-speaking Latvians accessed only Russian organic media. This was simply not the case, and very small, but nonetheless influential, organic Latvian Russian websites carried a far greater influence than originally assumed.[32]

Finally, social media has become ubiquitous, but it is possible to mistake volume for precision. With so much data available, it is increasingly important to find the right media, at the right time, and for the right audience, and this requires detailed research. A failure to understand any one of these points will make social media exploitation irrelevant and potentially misleading.

## SOCIAL MEDIA AND INFORMATION OPERATIONS (IO)

The second question we wish to address is how social media has changed the IO landscape. Joint Publication (JP) 3-13, *Information Operations* defines IO:

> as the integrated employment, during military operations, of Information Related Capabilities (IRC) in concert with other lines of operation to influence, disrupt, corrupt, or usurp the decision making of adversaries and potential adversaries while protecting our own.[33]

Given the impact on the information environment and therefore IRCs, undoubtedly, the answer to this is yes.

First and foremost, the ubiquity of social media has made it far easier for hostile elements to communicate directly with American audiences. Although the effectiveness of ISIS's online recruiting campaign is the subject of some conjecture—with some arguing that social media simply serves as a catalyst for other deeper behavioral drivers such as marginalization, poor life chances, or criminality—it is clear that imagery from the frontline, often of the most gruesome and violent nature, is immediately available to a population, which may or may not be directly influenced by it. Indeed, many argue that it is primarily through social media that extremist groups in the Middle East and Africa have been able to radicalize and recruit volunteers, in particular, to either conduct attacks in their own countries, or to travel to conflict areas and join the extremist organizations directly. One of the earliest exploiters of social media in this way was Al Shabbab in Somalia, whose delivery of high-quality videos through a range of platforms was at the time unprecedented.[34] ISIS later adopted and developed Al Shabbab's techniques.

It is believed that, so far, as many as 30,000 foreign fighters have traveled into Syria to take part in the conflict there, from countries all over the world. The reasons why they have traveled vary widely, with research showing that common motivations include excitement and adventure, peer pressure, and a search for identity. We argue that social media on its own is rarely the root cause; however, it facilitates radicalization and allows users to connect with ISIS recruiters as well as other radicalized individuals. Social media also allows for the creation of filtered information environments where the group dynamic can be a powerful radicalizing force.[35]

The use of social media by Russia in its ongoing operations in Eastern Ukraine is well known, and numerous papers have discussed it in detail.[36] While Russian operations in Crimea have received many internal Russian plaudits, and are the subject of numerous Russian military papers and studies, this is not a characteristic of operations in Eastern Ukraine. Russia's efforts there have been far less successful, partly due to being less planned and more reactive to events on the ground, but also arguably because it is significantly less disciplined. This is particularly the case in social media where the "firehose of nonsense and lies," a constant stream of trolling and misinformation, has been turned on, but seemingly with little effect other than for the West to constantly now question the veracity of any Russian communication. If it has been successful, it has been with the domestic Russian audience, which now is completely isolated from any competing narrative, giving the lie to the oft-stated concept of a global media space.

Russian operations in Donbass have also illustrated that social media usage can increasingly provide critical evidence of the reality of a situation in a combat environment. It is the ongoing position of the Russian government that Russian troops are not present in Eastern Ukraine. In common with other young people, globally, Russian troops are reminded to use social media to communicate with their loved ones and also with each other. Organizations such as Bellingcat have used this to track individual soldiers and the Russian units that command them. This has been particularly effective in building the case for Russian involvement in the shooting down of a Malaysian airliner.[37]

These are also risks that our own military could equally face. In a recent exercise presented at the NATO school in Oberammergau, one of this Letort Paper's authors demonstrated how the key personnel of a Royal Navy warship, their families, their social groups, their personal addresses, and their children's schools could all be easily identified from social media. In one case, the home address of a ship's Principal Warfare Officer was quickly found because he had applied for local authority planning permission, which is placed online in the UK, to extend his house. Possessing an unusual surname, he was easy enough to locate.[38] This type of information is almost impossible to protect and yet relatively easy to access using social media analysis tools. The Royal Navy ship in question may be one of the most powerful warships afloat, but without its key personnel present and focused on their tasks, it would have little operational utility. In a conflict environment, it would be naïve to think that our adversaries would not use this information, proactively, to their operational advantage. An integral part of Western IO doctrine is operational security (OPSEC); as well as regarding social media as an offensive operational tool, the Army needs to think very carefully about defensive measures in the social media domain. While we may be able to control our soldiers, controlling the use of social media by the soldiers' families will be significantly harder, if not impossible; and yet, through those very innocent conduits, it will be possible to directly target military personnel.

The continuing evolution of social media has undoubtedly resulted in areas of opportunity for friendly force IO as well. The nature of the Internet and social media in particular has effectively removed geographic barriers to communication with audiences across the world. Social media has none of the range or access limitations of radio or television. All that is required virtually anywhere in the world is Internet access. In addition, little infrastructure or resources are required to generate product. In many cases, little training has been required or conducted with adversarial groups rapidly learning "on the job." One unintended consequence of this accessibility of social media is that, whereas previous U.S. IO activity using conventional media has been able to be specifically targeted against adversarial groups or hostile governments, the use of social media in theory allows the U.S. home audience to access the same material. U.S. legislation states that U.S. forces cannot conduct IO against the U.S. population; however, clearly targeted material, even if able to be seen by a U.S. audience, would not violate this condition.[39] At the tactical level, we have seen how our adversaries have operationalized social media particularly in deceptive operations. For example, during the battle for Deir ez-Zour in Syria in 2014, between ISIS and al-Nusra, fake videos were released onto YouTube purporting to be tribal elders switching allegiance to ISIS.[40] A small number of defection videos were also released—that may have been false—causing al-Nusra to make public statements that they were not true, which then confused the rather sparse and poorly connected communities caught up in the fighting.

Deception, of course, is not new to the battlefield; but its dissemination on social media is, and it brings with it difficulties of our own making. Can social media, for example, be used in a tactical deception operation? Imagine simulating the movement of a patrol through the generation of social media reports of its progress. Under current interpretations of social media, to do so would be classified as "Black Psychological Operations (PSYOPS)," because it would be doctrinally categorized as a false report emanating from a false source. Both national and NATO Alliance doctrine specifically articulate that PSYOPS products should be truthful and attributable. Therefore, in our consideration of how the Army may leverage social media, a further area that needs to be considered is both doctrine and rules of engagement. It is perhaps worth noting that neither Operation FORTITUDE, the Second World War Allies' successful attempts to deceive the Nazis from knowing the location of the D-Day landings, nor Operation RHINO, General Norman Schwartzkopf's deception plan in the 1991 Gulf War, would be likely to succeed today given the omnipresence of social media. After all, the world first learned of the operation to kill Osama Bin Laden, while it was being undertaken, by a tweet from Pakistani IT consultant Sohaib Athar, who informed the world: "Helicopter hovering above Abbottabad at 1AM (is a rare event)."[41] Today, that might have been a live video feed from Periscope.

In fact, the whole issue of doctrine is problematic, because most of the doctrine that currently exists was written before the explosion in social media, or it only touched upon it very lightly or obliquely. Take for example the current official U.S. Army document that deals with social media: a handbook that was designed for PAO offices and primarily discusses social media in the context of engagement over news issues, rather than how it can be used on operations.[42]

The UK Army Doctrine Publication (ADP), *Operations,* describes itself as:

> the British Army capstone doctrine containing the enduring philosophy and principles for our approach to operations reflecting the rapidly involving dynamics of the contemporary operating environment.

It contains not one single reference to social media.[43] Similarly, the U.S. Joint Publication 1, *Doctrine for the Armed Forces of the United States*, describes itself as: "the capstone joint doctrine publication . . . for unified action by the Armed Forces of the United States"; yet, across some 172-pages, there is but one single mention of social media.[44] We should not really be surprised at this; doctrine, after all, takes a considerable amount of time to draft, be approved, and then published—all the more so in large and bureaucratic international organizations, such as NATO, where it inevitably becomes a reflection of the lowest common denominator of agreement between multiple member states. However, this does highlight the reality that the pace of technological and social change with social media is taking place faster than we can articulate in our military publications. This is, therefore, a significant challenge for the Army. How far do we wish as subordinates to innovate and take risk in a highly dynamic information environment? To succeed in future conflicts we may need to ask some very difficult questions about rules of engagement, legality, truthfulness, attribution, and a host of subsidiary issues, because future conflicts will be played out in a very different information environment.

It would be timely to consider the developments made by China in this particular area. In 2003, the Chinese Communist Party Central Committee signed off on a new concept called Three Warfares. They believed that in the future, nuclear arms would be unusable and that conventional kinetic options would be preferred in only a tiny number of possible future conflicts, and even then, outcomes would be problematic. Three Warfares is the adoption of a third type of warfare, where PSYOPS, media manipulation, and legal warfare are the mainstays of Chinese strategy. We have seen this played out in key areas of Chinese foreign policy in the present-day, for example, with the physical creation of new islands in disputed areas of the South China Sea that has provided China with opportunities to "legally" extend its territorial boundaries. In 2015, China created a strategic support force commanded by a four-star general. Within this force, China placed both its defensive and offensive cyber PSYOPS and media operations commands.

What we might presume from this is that both Russia and China are putting great emphasis on operationalizing emerging technologies such as social media. This does not yet appear to be the case for the U.S. Army, and we believe there are a number of key areas that need to be investigated.

**Force Structures, Training & Education, and Equipment.**

Given the nature and pace of social media development, it is clear to us that the U.S. Army must address three distinct issues: Force Structures; Training and Education; and Equipment. This Letort Paper and its two English authors would not propose to recommend specific force structures, but we could generically recommend that a social media capacity needs to be built into every level of command. In the real-time information environment created by social media, operational commanders at every level will need simultaneous and identical situational awareness of unfolding events. It is preposterous to presume that this increasingly vital area should somehow reside only with the PAO and Intelligence (INT); it must be central to operations and planning. Today, it is entirely normal to have a live unmanned aerial vehicle (UAV) feed into the Force Operations room; in the future, social media feeds should be similarly entirely normal.

*Equipment.* There is a critical need to integrate the Internet into operational headquarters at all levels of command, in particular mobile and tactical headquarters.[45] Without broad access across the staff functions, not just IO, it will not be possible to properly address a number of the issues identified earlier. While there are technical and security issues with ensuring Internet access within headquarters and tactical headquarters in particular, overall the issue is more about integrating existing technology rather than the challenge of developing new technology.

*Analytical software tools.* A great deal of emphasis has been placed on the development of analytical software for social media analysis. In the context of IO, both their importance and their limitations need to be properly recognized. There are now large numbers of commercial applications that scrape open source media and provide all kinds of metrics. Such analysis can be very useful, but there is an urgent need to understand the different requirements that different staff functions might have. Both INT and PAO functions and requirements are relatively well understood, but the requirements for others, including both Operations and IO are less clear.

Social media needs to be considered in terms other than simply as a media channel, and as a result operational headquarters at all levels require a monitoring function that reflects this. A social media feed, suitably filtered and managed appropriately for the level of headquarters, needs to be available to the INT staff function and sufficiently integrated with the Operations, IO, and PAO staff functions. This function will need to be capable of live monitoring, but depending on the level of headquarters may only need limited analysis capability, as depth analysis of social media can be conducted with reachback capabilities. Responsibilities for the different staff functions need to be clarified and manpower changes may be required as a result of the need for new social media roles within headquarters and supporting IO and PSYOPS components.

For example, in the future it is highly likely that if deployed troops encounter civil disobedience, including demonstrations and riots, these will be directed and coordinated through social media in real time, possibly also with multiple sources broadcasting the events on live streaming video directly onto the Internet. Under these circumstances, a capability for the operations staff to monitor events on social media directly would be required in the same way that a UAV feed can be monitored directly when necessary. This is not revolutionary, it is evolutionary; and in essence, this is no different to the existing practice of monitoring radio communications, already updated in recent operational contexts to the real-time provision of tactical information by listening in to Integrated Communications Security (ICOM) chatter. As those involved in such civil disobedience would be a key audience for IO, there is also both a monitoring implication and an IO opportunity for intervention, which implies that IO needs a significant deployable social media broadcast capability that has a capacity to operate in real time as well as a longer-term engagement capacity.

Effective and timely communication in a rising number of areas across the world will require engagement via social media as they increasingly come to dominate how local audiences communicate and receive information. IO and PSYOPS capabilities will need sufficient capability to broadcast using different social media platforms. Specific levels of capacity will require a more detailed analysis of the likely tasks at each level of command, and there is also a need to consider different forms of reachback capability. Nevertheless, in line with current U.S. doctrine, commanders will need to retain some level of capability at different levels of command to be able to utilize IO in direct support of their own operations.

*Training and Education.* We believe there exists a requirement for all personnel to be better educated in terms of understanding the full range of functions that social media can be used for; in particular, dispelling the myth that only PAO or INT need be aware of or understand social media. This basic education level would also include a clear understanding of the OPSEC risk to U.S. forces that social media represents. There is a need for commanders, and in particular operations and INT staff, to understand social media and its different functions, not simply its media function, in order to understand its potential impact on operations and to incorporate that understanding into their planning and operations. Commanders do not need specialist social media knowledge, but they do need sufficient knowledge to understand its importance, as well as how the different stakeholders in conflict environments are likely to employ it. They also need to have an understanding of the analytical techniques that can be used on social media data and the capabilities and limitations of that analysis.

Specialist IO personnel will need to have a greater depth of understanding of social media, of its functions, and of the psychology underpinning human behavior online. They also need an understanding of the different types of analysis that can be conducted on social media data and the relative value of that analysis.

IO social media analysts will increasingly be required. The huge increase in social media use presents increasing opportunities to mine data-rich information in a conflict environment and collect raw intelligence on different actors and stakeholders within them. This has the potential to generate detailed assessments and can offer deep insights into audience motivations, attitudes, and behaviors. In addition to requiring specialist analytical skills, these analysts need to be able to integrate such information with other target audience data and intelligence to best support IO.

In the UK, the use of competency frameworks has become increasingly useful in defining who needs to know what and to what detail. We would suggest a competency framework for social media (see Table 1).

| Competency | Audience |
|---|---|
| Awareness<br>Understand the threats and opportunities, utility and constraints of social media. | All |
| Working Familiarity<br>Can access social media for analysis purposes and has a deeper understanding of its utility and place in intelligence collection and operational planning. | G2-G5 |
| Practitioner<br>Can access social media for analysis and influence purposes and has a deep understanding of its utility and intelligence collection and operational planning. | PAO/ INT/ IO |
| Expert<br>Takes primary responsibility for all aspects of social media usage. | IO |

**Table 1. Competency Framework for Social Media.**

*Technology*. The technical challenges involved in integrating Internet access with classified systems center on the security challenges of doing so, rather than the development of new technology. In contrast, many of the technologies involved in the area of big data analysis are new and emerging, with some important areas such as sentiment analysis still in a relatively early stage of their development.[46] Further developments in these areas will only increase the value and importance of the analysis of social media in conflict environments and reinforce the need for such analysis to be provided in a timely manner to operational commanders at all levels. Finally, language translation technology continues to evolve, and it increasingly will enable operators to utilize social media in a multi-language environment.[47]

## CONCLUSION

The impact of social media on the media environment has been widely recognized; as has the ability of extremist and adversarial organizations to exploit social media to publicize their cause, spread their propaganda, and recruit vulnerable individuals. However, the full implications of the mobile and social media revolution are not yet fully understood, and they extend far beyond these areas. Social media will increasingly have a direct impact on virtually all aspects of military operations in the 21st century, through all four of the different applications of social media we have described. In doing so, social media will force significant changes to policy, doctrine, force structures, and virtually all staff functions within operational units. Social media will also necessitate new training requirements and new approaches to traditional operational challenges. All this will require a wider understanding of the full impact of social media and the realization that it is no longer an area simply of concern to PAO and possibly INT.

Although, with new threats come new opportunities, and the interactive nature of social media in particular means that it is potentially a very powerful medium for IO, if that potential is recognized and capabilities can be developed to respond to this rapidly developing revolution.

## POLICY RECOMMENDATIONS

In addition to the general recommendations above, we make the following specific proposals:

1. Social media is integral to the conduct of operations, not just a PAO or INT function. Thus, the U.S. Army **needs a custom-made doctrine** and an **educational capacity** to inform commanders on its intelligent and safe use.
2. In order to integrate social media into operations, each of the **staff functions needs continuous access** to the Internet at their desk.
3. **Social media must be integral to all exercises.** This needs the creation of a virtual social media environment—i.e., a sandbox or simulation—to add the social media space to training environments. This must be a space in which it is "safe to fail."
4. There is a **need for commanders, in general, and operations and INT staff, in particular, to understand social media and its different functions,** not limited to media use, and **how to incorporate it into the planning cycle.** All staff functions need training and education, possibly using a competency framework.
5. Within headquarters, the **responsibility for social media needs to be clearly articulated** and the posts resourced appropriately.
6. U.S. Army and Department of Defense (DoD) **policy and doctrine must clarify a host of difficult issues relating to social media usage** and, in particular, its use in deception and PSYOPS.

## ENDNOTES

1. "Most famous social network sites worldwide as of January 2017, ranked by number of active users (in millions)," Statista.com, January 2017, available from *www.statista.com/statistics/272014/globalsocialnetworks-ranked-by-number-of-users/*.

2. Jacob Poushter, "Smartphone Ownership and Internet Usage Continues to Climb in Emerging Economies: But advanced economies still have higher rates of technology use," *Global Technology Report*, Washington, DC: Pew Research Center, February 22, 2016, available from *www.pewglobal.org/2016/02/22/smartphone-ownership-and-internet-usage-continues-to-climb-in-emerging-economies/*, accessed February 23, 2016.

3. "Cisco Visual Networking Index: Global Mobile Data Traffic Forecast Update, 2015–2020," *White Paper*, San Jose, CA: Cisco Systems, Incorporated, updated June 1, 2016.

4. Serene Assir, "Facebook, WhatsApp and Viber light way to Europe for Syrian refugees," *The Times of Israel*, August 19, 2015, available from *www.timesofisrael.com/facebook-whatsapp-and-viber-light-way-to-europe-for-syrian-refugees/*.

5. Annie Shiel, "Conflict Crowdsourcing: Harnessing the Power of Crowdsourcing for Organizations Working in Conflict and the Potential for a Crowd Sourced Portal for Conflict-Related Information," *Undergraduate Honours Thesis*, Montréal, Québec, CA: McGill University, December 3, 2013.

6. Jenny Hauser, "Speed in Context: Real-time News Reporting and Social Media," Paper Presented at the European Journalism Training Association Conference held at the Dublin Institute of Technology, School of Media, on October 24, 2014.

7. The British Broadcasting Corporation's (BBC) *Editorial Guidelines*, state that:

> [the BBC will] gather material using first hand sources wherever possible, check and cross check facts, validate the authenticity of documentary evidence and digital material, corroborate claims and allegations made by contribu-

tors wherever possible. In news and current affairs content, achieving due accuracy is more important than speed.

British Broadcasting Corporation, *Editorial Guidelines*, London, UK: British Broadcasting Corporation, 2017, available from *www.bbc.co.uk/editorialguidelines/guidelines/accuracy*.

8. The main U.S. Army official publication that refers to social media is U.S. Department of the Army, *The United States Army Social Media Handbook*, Washington, DC: U.S. Army, Office of the Chief of Public Affairs, Online and Social Media Division, April 2016, available from *https://www.army.mil/e2/rv5_downloads/socialmedia/army_social_media_handbook.pdf*, which does so in the context of public affairs.

9. For example, one has only to look at the volume of information created online by the Internet Research Agency based at 55 Savushkina Street, St. Petersburg, Russia in support of events such as the downing of Malaysia Airlines Flight 17 (MH17).

10. Associated Press (AP), "Facebook, Twitter, WhatsApp helping ISIS, U.K. spy alleges," CBC News, November 4, 2014, available from *www.cbc.ca/news/world/facebook-twitter-whatsapp-helping-isis-u-k-spy-alleges-1.2822890*.

11. Megan Specia, "WhatsApp offers lifeline for Syrian refugees on journey across Europe," Mashable, July 3, 2015, available from *mashable.com/2015/07/03/syrians-europe-whatsapp-refugees/#ITbIBHPBFOq8*.

12. Alina Selyukh and Camila Domonoske, "The Apple-FBI Debate Over Encryption: Apple, The FBI And iPhone Encryption: A Look At What's At Stake," The Two-Way: Breaking News From NPR, National Public Radio (NPR), February 17, 2016, available from *www.npr.org/sections/thetwo-way/2016/02/17/467096705/apple-the-fbi-and-iphone-encryption-a-look-at-whats-at-stake*.

13. Russ Linden, "The Life-Saving Power of Crowdsourcing," *Governing: The States and Localities*, Management Insights, January 23, 2013, available from *www.governing.com/columns/mgmt-insights/col-crowdwourcing-ushahidi-saving-lives-haiti-earthquake.html*.

14. *Ibid.*

15. Sean Aday, Henry Farrell, Marc Lynch, John Sides, and Deen Freelon, "Blogs and Bullets II: New Media and Conflict After the Arab Spring," *Peaceworks*, No. 80, Washington, DC: United States Institute of Peace, July 2012, available from *www.usip.org/publications/blogs-and-bullets-ii-new-media-and-conflict-after-the-arab-spring*.

16. Mark P. Mills, "The Mobile Revolution Has Only Just Begun," *Forbes*, January 19, 2015, available from *www.forbes.com/sites/markpmills/2015/01/19/the-mobile-revolution-has-only-just-begun/#7e481fd9904c*.

17. "Cisco Visual Networking Index."

18. *Ibid.*

19. *Ibid.*

20. Simon Kemp, "Digital in 2016," *Special Reports*, Singapore: We Are Social Limited, January 27, 2016, available from *wearesocial.com/uk/special-reports/digital-in-2016*.

21. *Ibid.*

22. Damian Radcliffe, "Survey: Arab Youth consume less news and trust social media as a news source," Knowledge Bridge, June 27, 2013, available from *www.kbridge.org/en/arab-youth-are-consuming-less-news-and-increasingly-trust-social-media-as-a-news-source/*.

23. Ammar Halabi, "The use of social media in Syria," Social Informatics Blog, April 24, 2014, available from *https://socialinfoblog.wordpress.com/2014/04/*.

24. Mark Hoelzel, "Update: A breakdown of the demographics for each of the different social networks," *Business Insider UK*, June 29, 2015, available from *uk.businessinsider.com/update-a-breakdown-of-the-demographics-for-each-of-the-different-social-networks-2015-6?r=US&IR=T*; Verto Analytics, "The Demographics of Social Media Properties: Looking Beyond Downloads," Consumer Insights, October 21, 2015, available from *www.vertoanalytics.*

*com/2015/10/the-demographics-of-social-media-properties-looking-beyond-downloads/*, accessed June 1, 2016.

25. "Draugiem," crunchbase.com, n.d., available from *www.crunchbase.com/company/draugiem-2*, accessed June 1, 2016; "Top sites in Latvia," Alexa.com, n.d., available from *www.alexa.com/topsites/countries/LV*, accessed June 1, 2016; A target audience analysis (TAA) was undertaken by the London, UK-based company, Strategic Communication Laboratories Limited in 2015.

26. Carol Matlack, "The Kremlin Tried to Use VKontakte—Russia's Facebook—to Spy on Ukrainians," *Bloomberg Businessweek*, April 17, 2014.

27. Khalid Albaih, "How WhatsApp is fuelling a 'sharing revolution' in Sudan," *The Guardian*, October 15, 2015, available from *www.theguardian.com/world/2015/oct/15/sudan-whatsapp-sharing-revolution.*

28. Internet World Stats website, "Asia: Asia Marketing Research, Internet Usage, Population Statistics and Facebook Information," Afghanistan, last updated February 16, 2017, available from *www.internetworldstats.com/asia.htm#af.*

29. Mirwais Khan, "Helmand governor takes to Facebook to warn Afghan president of Taliban threat," *Independent*, December 20, 2015, available from *www.independent.co.uk/news/world/asia/helmand-governors-taliban-warning-on-facebook-a6780846.html.*

30. Sam Schechner, "Erdogan Embraces Social Media to Repel Coup Attempt in U-Turn," *The Wall Street Journal*, July 17, 2016, available from *https://www.wsj.com/articles/erdogan-embraces-social-media-to-repel-coup-attempt-in-u-turn-1468760698.*

31. Greg Miller, "National Security: Panel casts doubt on U.S. propaganda efforts against ISIS," *The Washington Post*, December 2, 2015, available from *https://www.washingtonpost.com/world/national-security/panel-casts-doubt-on-us-propaganda-efforts-against-isis/2015/12/02/ab7f9a14-9851-11e5-94f0-9eeaff906ef3_story.html.*

32. Target audience analysis (TAA) was undertaken by London, UK-based company, Strategic Communication Laboratories Limited in 2015.

33. U.S. Joint Chiefs of Staff, *Information Operations*, Joint Publication 3-13, Washington, DC: U.S. Joint Chiefs of Staff, November 27, 2012, Incorp. Change 1, November 20, 2014.

34. Will Oremus, "Twitter of Terror: Somalia's al-Shabaab unveils a new social media strategy for militants," Slate, December 23, 2011, available from *www.slate.com/articles/technology/technocracy/2011/12/al_shabaab_twitter_a_somali_militant_group_unveils_a_new_social_media_strategy_for_terrorists_.html.*

35. Research interview conducted by the authors with Hanif Qadir, Chief Executive Officer of The Active Change Foundation on May 16, 2016.

36. For example, see North Atlantic Treaty Organization (NATO) Centre of Excellence Strategic Communications Centre of Excellence, *Framing of the Ukraine–Russia Conflict in Online and Social Media*, Riga: North Atlantic Treaty Organization (NATO) Centre of Excellence Strategic Communications Centre of Excellence, May 2016, available from *www.stratcomcoe.org/framing-ukraine-russia-conflict-online-and-social-media.*

37. Daniel Romein, "MH17—Potential Suspects and Witnesses from the 53rd Anti-Aircraft Missile Brigade: A Bellingcat Investigation," Bellingcat.com, February 23, 2016, available from *https://www.bellingcat.com/news/uk-and-europe/2016/02/23/53rd-report-en/.*

38. "Using Social Media for Operational Effect," A lesson delivered to the NATO School Oberammergau Senior Officer's Information Operations Course in May 2016.

39. The *United States Information and Educational Exchange Act of 1948*, Public Law 80-402, Washington, DC: U.S. Government Printing Office, 1948, popularly referred to as the Smith–Mundt Act, specifies the terms in which the U.S. Government can engage global audiences.

40. Valerie Szybala, "The Islamic State of Iraq and al-Sham and the 'Cleansing' of Deir ez-Zour," *Backgrounder*, Washington, DC: Institute for the Study of War, May 14, 2014, available from *www.understandingwar.org/backgrounder/islamic-state-iraq-and-al-sham-and-%E2%80%9Ccleansing%E2%80%9D-deir-ez-zour.*

41. Sohaib Athar (@ReallyVirtual), "Helicopter hovering above Abbottabad at 1AM (is a rare event)," May 1, 2011, 12:58 p.m., tweet, available from *https://twitter.com/ReallyVirtual?lang=en-gb.*

42. U.S. Department of the Army.

43. UK Ministry of Defence (MOD), *Operations*, Army Doctrine Publication (ADP), Shrivenham, UK: Development, Concepts and Doctrine Centre, Ministry of Defence, November 2010.

44. U.S. Joint Chiefs of Staff, *Doctrine for the Armed Forces of the United States*, Joint Publication 1, Washington, DC: U.S. Joint Chiefs of Staff, March 25, 2013, pp. i., I-12.

45. From our joint experience of operational headquarters, we know this to be the case; however, one may argue that while there is no doctrine that tackles this issue there is no incentive to remedy this situation. Our experience suggests that commanders will continue to view social media purely as a potential breach of operational security (OPSEC) and, therefore, limit its use as much as possible.

46. Matthew Mooney, "Facebook Reactions: The Future for Sentiment Analysis," Isitsweet.wordpress.com, November 17, 2015.

47. Quentin Hardy, "Language Translation Tech Starts to Deliver on Its Promise," Bits, blog of *The New York Times*, January 11, 2015, available from *https://bits.blogs.nytimes.com/2015/01/11/language-translation-tech-starting-to-deliver-on-its-promise/?_r=0.*

U.S. ARMY WAR COLLEGE

Major General William E. Rapp
Commandant

*****

STRATEGIC STUDIES INSTITUTE
and
U.S. ARMY WAR COLLEGE PRESS

Director
Professor Douglas C. Lovelace, Jr.

Director of Research
Dr. Steven K. Metz

Authors
Mr. Ian Tunnicliffe
Dr. Steve Tatham

Editor for Production
Dr. James G. Pierce

Publications Assistant
Ms. Denise J. Kersting

*****

Composition
Mrs. Jennifer E. Nevil

Made in the USA
Monee, IL
04 November 2026

67376788R00030